Black Holes

1, Volume 1

Zac Apos

Published by Ange Antony, 2025.

BLACK HOLES

First edition. September 24, 2025.

ISBN: 979-8230636014

Written by Zac Apos.

Table of Contents

To the curious minds who look up at the night sky and wonder — this is for you. May this book ignite your imagination, deepen your sense of wonder, and inspire you to explore the mysteries of the universe that lie beyond the stars.

We hope you enjoy this journey — stay tuned, because the story of black holes is just beginning.

BLACK HOLES

Gateways to the Unknown

Introduction

Black holes are among the most mysterious and fascinating objects in the universe. They are regions of space where gravity is so powerful that not even light—the fastest thing in existence—can escape. For centuries, humanity has gazed up at the night sky, wondering what lies beyond the stars, but black holes represent an even deeper mystery: what lies beyond the very fabric of space and time itself?

The concept of black holes is both terrifying and captivating. On one hand, they are cosmic monsters, swallowing entire stars that wander too close. On the other, they are essential pieces of the cosmic puzzle, helping scientists understand gravity, matter, energy, and even the origins of the universe. They are not just destructive—they are also creative, playing a key role in shaping galaxies and influencing the evolution of the cosmos.

Although black holes sound like science fiction, they are very real. Modern telescopes, satellites, and observatories have confirmed their presence—not through direct images, but through their powerful effects on the space around them. From the discovery of quasars in the mid-20th century to the first-ever image of a black hole in 2019, the journey of uncovering these cosmic giants is one of the greatest adventures in science.

This book will take you through that journey. We will start with the origins of the black hole idea, move through the anatomy of these strange objects, explore how they are formed, learn how scientists detect them, and see how they shape the universe around them. Along the way, we will dive into concepts of time, relativity, and even the ultimate fate of the universe.

Black holes challenge everything we think we know about reality. They blur the line between physics and philosophy, between the known and the unknown. By the end of this book, you will see why black holes regions of darkness in space are not just, but gateways to understanding the deepest laws of the universe.

Chapter 1: The Birth of an Idea

The story of black holes begins not with modern physics, but centuries ago with humanity's first attempts to understand gravity and the forces that govern the cosmos. It is a tale that stretches from Newton's revolutionary equations to the cutting-edge theories of today.

Newton's Revolution

In the late 1600s, Sir Isaac Newton formulated his law of universal gravitation, arguably one of the most important breakthroughs in the history of science. Newton demonstrated that every object in the universe attracts every other object with a force proportional to their masses and inversely proportional to the square of the distance between them. This was not just a mathematical equation—it was a profound shift in how humanity understood the cosmos.

Newton's insights explained phenomena that had puzzled scientists for centuries: why apples fall to the ground, why the Moon orbits the Earth, and why planets revolve around the Sun in elliptical paths. His work created a framework for celestial mechanics that endured for more than two centuries, and became the foundation for all classical physics.

But even in Newton's time, some thinkers began to imagine far stranger possibilities. What if there existed a body so dense and massive that its gravitational pull would be so strong that even light could not escape?

Michell and the "Dark Star"

In 1784, John Michell, an English natural philosopher and clergyman, ventured this exact idea. In a paper presented to the *Philosophical Transactions of the Royal Society*, Michell proposed the concept of a "dark star." His reasoning was simple yet profound: if a star had enough mass and density, its escape velocity—the speed necessary to break free from its gravitational pull—would exceed the speed of light. Since nothing can travel faster than light, such an object would be invisible.

Michell even suggested ways such dark stars could be detected indirectly, by studying the motions of nearby objects. His ideas were remarkably prescient—centuries ahead of experimental confirmation—but they remained largely speculative at the time.

Laplace and Early Speculations

Around the same time, Pierre-Simon Laplace, the French mathematician and astronomer, also considered the possibility of invisible, gravitating bodies. In the first edition of his *Exposition du Système du Monde* (1796), Laplace mentioned objects with gravity so strong that light could not escape. However, by the second edition, he removed this reference. Why? By then, the nature of light had been debated, and many scientists believed light to be a particle. Later, with the wave theory of light gaining acceptance, the idea of dark stars was set aside for more than a century.

Einstein's Revolution

The story might have ended there if not for Albert Einstein. In 1915, Einstein published his General Theory of Relativity, fundamentally altering our understanding of gravity. Newton had described gravity as a force between masses, but Einstein showed it was far more subtle and profound: gravity is the warping of spacetime itself. Massive objects curve spacetime, and smaller objects travel along those curves. This was a revolutionary idea—gravity was no longer just a force, but a manifestation of the geometry of the universe.

Einstein's equations predicted the possibility of extreme distortions of spacetime—objects so dense they would form singularities. But Einstein himself doubted their physical reality, seeing them as mathematical curiosities rather than actual celestial objects.

Schwarzschild's Solution

Shortly after Einstein published his theory, a German physicist named Karl Schwarzschild found a solution to Einstein's equations that described such a singularity. Remarkably, he did this while serving on the Russian front during World War I, under unimaginable conditions. Schwarzschild's solution described a point in space where gravitational pull becomes infinite and no escape is possible—even for light. This was the first real mathematical description of what we now call a black hole.

From "Frozen Stars" to Black Holes

For decades, the notion of black holes was treated as theoretical speculation. The term itself did not exist until the mid-20th century. In the 1960s, physicist John Wheeler coined the term "black hole," replacing older terms like "frozen star" or "collapsar." Wheeler's evocative phrase captured the public imagination and cemented black holes as central to modern physics.

By then, new observational astronomy techniques were providing evidence of strange X-ray sources and high-energy phenomena consistent with black holes. Work by scientists such as Roger Penrose, Stephen Hawking, and Yakov Zel'dovich during the 1960s and 70s deepened understanding, showing that black holes were not just theoretical but real—and far stranger than anyone imagined.

A Journey of Discovery

From Michell's "dark star" to Wheeler's black hole, the concept evolved over centuries. It is a story of human curiosity, persistence, and imagination—a story of how ideas once dismissed as impossible have become central to modern physics. The birth of the black hole idea is a testament to the power of bold thought, rigorous mathematics, and relentless observation.

Chapter 2: Anatomy of a Black Hole

A black hole might seem like a simple void—an empty, dark pit in space—but in reality, it is a far more intricate and fascinating object. What lies within a black hole is the subject of some of the most profound questions in physics, bridging the limits of our knowledge. While we cannot see a black hole directly, astronomers and physicists have developed a surprisingly detailed map of their structure. This knowledge comes from a combination of mathematical theory, astronomical observation, computer simulations, and decades of exploration into the extremes of the universe.

The Event Horizon: The Point of No Return

The most defining feature of a black hole is its event horizon. This is not a physical surface like the crust of a planet, but an invisible boundary marking the limits of escape. Once anything—whether a particle, a photon of light, or a spaceship—crosses the event horizon, there is no return. The escape velocity inside this boundary exceeds the speed of light, making it impossible for any information to escape.

The event horizon is the ultimate frontier in physics. From the outside, objects falling toward a black hole appear to slow down, stretching out in time, as if frozen at the brink of oblivion. This curious effect is a direct consequence of Einstein's theory of general relativity, where gravity distorts time itself. To a distant observer, the process of falling in takes an infinite amount of time. Yet to the object itself, crossing the horizon happens in an instant.

This strange dichotomy of perspectives was beautifully illustrated in the 2014 film *Interstellar*, where the effects of time dilation near a black hole were depicted with scientific rigor, consulting physicist Kip Thorne, a Nobel laureate. This blend of theory and cultural depiction shows just how tantalising the event horizon is—it is both a real boundary in the cosmos and a gateway to the unknown.

The Singularity: Where Physics Breaks Down

At the very heart of every black hole lies the singularity. It is a point of infinite density where the gravitational pull becomes so intense that spacetime itself curves infinitely. In theory, all matter that falls into a black hole is crushed into this point, where known laws of physics no longer apply.

The singularity is not just a puzzle—it is a challenge to the foundations of science. Stephen Hawking, Roger Penrose, and other physicists have wrestled with the implications for decades. Penrose's work on singularities, which earned him a Nobel Prize in 2020, demonstrated that singularities are not merely artifacts of idealised mathematics, but inevitable consequences of general relativity under certain conditions. Hawking's work on black hole thermodynamics and quantum effects added another layer of mystery, suggesting that even singularities might not be the final word on what lies inside a black hole.

Some modern theories suggest that singularities could be replaced by exotic structures predicted by quantum gravity. Ideas such as "fuzzballs" in string theory or loop quantum gravity's "quantum bounce" hint that spacetime might be far more intricate inside a black hole than we imagine. These remain speculative but highlight how black holes push physics to its absolute limits.

The Accretion Disk: Cosmic Light Shows

While black holes themselves emit no light, they often produce dazzling displays of energy and radiation. Surrounding many black holes is an accretion disk—a glowing ring of superheated gas, dust, and stellar debris spiralling inward. This matter accelerates under gravity, colliding and heating up to millions of degrees, emitting powerful radiation across the electromagnetic spectrum, including X-rays and gamma rays.

These emissions are not just scientific curiosities—they are how we know black holes exist. In the 1960s, the discovery of quasars (quasi-stellar objects) shocked astronomers. These incredibly bright sources of light, billions of light-years away, were later explained as accretion disks around supermassive black holes at galactic centres. Quasars remain some of the most luminous objects in the universe, shining with the energy of trillions of suns.

Accretion disks themselves are complex structures. Their dynamics involve magnetohydrodynamics, turbulence, and relativistic effects. Computer simulations now allow scientists to model these processes in remarkable detail, revealing jets, winds, and instabilities that paint a vivid picture of matter at the brink of oblivion.

The Ergosphere: Frame Dragging and Energy Extraction

In rotating black holes—known as Kerr black holes—there is another strange and fascinating region called the ergosphere. Located just outside the event horizon, the ergosphere is a distorted region of spacetime dragged along by the black hole's rotation. This effect, called frame dragging, means that nothing can remain stationary within this zone.

The ergosphere is not just a curiosity—it offers a tantalising possibility: extracting energy from a black hole. Physicist Roger Penrose proposed the "Penrose process," in which particles entering the ergosphere could split, with one falling into the black hole and the other escaping with more energy than it entered. This theoretical concept has inspired decades of research and remains a key idea in understanding black hole energetics.

Hawking Radiation: Black Holes Can Die

Contrary to the idea of black holes as eternal prisons, Stephen Hawking shocked the world in 1974 by showing that black holes are not entirely black. Using quantum mechanics, he demonstrated that particle-antiparticle pairs forming near the event horizon could result in the emission of radiation. This "Hawking radiation" implies that black holes can slowly lose mass and, over unimaginable timescales, evaporate entirely.

While Hawking radiation is negligible for massive black holes, its implications are profound. It links quantum mechanics, thermodynamics, and relativity, suggesting that black holes obey laws of entropy and information conservation. This insight opened a new field of research—black hole thermodynamics—which remains at the frontier of theoretical physics today.

The Beauty and Terror of Black Hole Anatomy

Altogether, the anatomy of a black hole reveals a world where physics is pushed to extremes. From the invisible veil of the event horizon to the crushing singularity, from blazing accretion disks to exotic ergospheres, black holes are cosmic laboratories for exploring the boundaries of reality.

They are not merely objects to be feared, they are gateways to understanding the universe itself, places where science confronts the limits of human knowledge. As we continue to study them, we are learning not just about black holes, but about the fundamental nature of space, time, and existence.

Chapter 3: The Formation of Black Holes

Black holes are not born out of nothing, they are forged in some of the most violent and transformative events in the universe. Their formation is a process that bridges the life cycle of stars, the evolution of galaxies, and the conditions of the early cosmos. To understand how black holes form, we must explore the life and death of stars, the physics of gravitational collapse, and some of the most extraordinary discoveries in modern astronomy.

Stars: Cosmic Furnaces of Creation

Stars are colossal engines of nuclear fusion. At their cores, hydrogen atoms combine under extreme pressure and temperature to form helium, releasing enormous amounts of energy in the process. This fusion creates an outward pressure that counteracts the inward pull of gravity, keeping the star stable for millions or billions of years.

The life of a star is a balance—a cosmic tug-of-war between gravity and nuclear fusion. For smaller stars like our Sun, this balance lasts for billions of years. When the fuel runs out, the star gradually expands into a red giant before shedding its outer layers to form a planetary nebula, leaving behind a dense core known as a white dwarf.

For massive stars—those many times the mass of the Sun—the ending is far more dramatic. When nuclear fusion can no longer sustain them, gravity overwhelms the outward pressure. The core collapses catastrophically, and the outer layers explode in a spectacular supernova, one of the most energetic events in the universe.

From Collapse to Singularity

In certain cases, the core left behind after a supernova has enough mass—generally greater than about three times the Sun's mass—that not even the pressure from neutron degeneracy can halt collapse. The core continues shrinking under its own gravity until it forms a singularity, an infinitely dense point where space and time break down, surrounded by an event horizon. A black hole is born.

One of the most dramatic confirmations of this process came from observations of the binary system Cygnus X-1, discovered in the 1960s. Cygnus X-1 was one of the first strong candidates for a stellar-mass black hole, with evidence showing a massive, unseen object pulling material from its companion star. This discovery marked a turning point in proving that black holes were not just theoretical constructs.

Primordial Black Holes: Relics of the Early Universe

While stellar collapse explains many black holes, it may not explain them all. Physicists have theorised the existence of primordial black holes—objects that may have formed in the first moments after the Big Bang. These would originate from regions of extreme density and curvature in the early universe, collapsing directly into black holes before stars even existed.

Primordial black holes remain speculative, but they could have profound implications. Some researchers suggest they could account for some or all of the universe's mysterious dark matter. Others propose they could have seeded the growth of supermassive black holes. Detecting them is a challenge, but projects such as gravitational wave observatories and surveys for unusual cosmic microlensing events are actively searching for clues.

Supermassive Black Holes: Giants of the Cosmos

The largest black holes in the universe, supermassive black holes—lie at the hearts of galaxies, including our own Milky Way. These giants contain millions to billions of times the mass of the Sun. Sagittarius A*, the supermassive black hole at the centre of our galaxy, has a mass of about four million suns and influences the motion of thousands of stars around it.

How supermassive black holes form is one of the great mysteries of astrophysics. Some theories propose that they grew over billions of years by consuming stars and gas, merging with other black holes. Others suggest they might have formed directly from the collapse of massive gas clouds in the early universe, bypassing the star stage entirely. The truth may involve a combination of processes.

The discovery of quasars in the mid-20th century—extremely luminous galactic nuclei powered by supermassive black holes—provided compelling evidence for their existence. Quasars, some visible across billions of light-years, demonstrate the enormous energy these objects can unleash.

Black Hole Mergers: Cosmic Collisions

Another dramatic pathway for black hole formation is through mergers. When two black holes orbit each other and eventually collide, they create a single, more massive black hole, releasing tremendous amounts of energy as gravitational waves.

The Laser Interferometer Gravitational-Wave Observatory (LIGO) confirmed this process in 2015 with the historic detection of gravitational waves from the merger of two black holes. This not only confirmed a century-old prediction of Einstein's theory but opened a new way to observe black holes, enabling scientists to "listen" to the universe in a fundamentally new way.

These observations revealed black holes with masses that were unexpected, challenging previous ideas about their formation and sparking fresh questions about the diversity and origins of black holes in the universe.

Galactic Evolution and Black Hole Formation

Black hole formation is intimately linked to galaxy evolution. As black holes grow by consuming matter, they influence the environment around them. The powerful jets and winds produced by supermassive black holes can regulate star formation in their host galaxies. In some cases, they can trigger new star formation; in others, they can blow away gas and suppress it. This delicate balance plays a key role in shaping the structure of galaxies across cosmic time.

Observations from instruments like the Event Horizon Telescope and space-based observatories such as the Chandra X-ray Observatory continue to reveal surprising details about this relationship. The discovery that nearly every large galaxy harbours a supermassive black hole underscores their importance in the cosmic story.

From Death Comes Creation

The formation of black holes is a story of transformation. From the death throes of massive stars to the turbulent conditions of the early universe, black holes emerge as both destroyers and creators. They mark the end of one cosmic chapter and the beginning of another.

In a sense, black holes are the universe's ultimate recycling centres—regions where matter and energy are transformed, reshaped, and returned to the cosmos in new forms. They are not just objects of destruction but agents of change, helping sculpt the cosmic landscape.

As we look deeper into space, studying black holes is not merely about understanding their origins—it is about understanding the origins of the universe itself. Every black hole tells a story, and each story is a chapter in the grand narrative of creation and destruction that shapes everything we see in the cosmos.

Chapter 4: Types of Black Holes

Black holes are far from uniform. They vary enormously in size, origin, and behaviour, and scientists classify them according to their mass, formation history, and the cosmic environments they inhabit. Understanding the different types of black holes is essential to grasping their role in the universe, and it also reveals how much diversity exists in these mysterious objects.

Stellar-Mass Black Holes: The Commonborn Giants

Stellar-mass black holes are the most common type. They form from the collapse of massive stars at the end of their life cycle, typically weighing between about three and tens of times the mass of our Sun. The size of a stellar-mass black hole is proportional to its mass, but even the largest are compact enough to fit within a sphere just a few dozen kilometres in diameter.

One of the first candidates for a stellar-mass black hole, Cygnus X-1, was discovered in the 1960s. This binary system features a visible star and an unseen companion, with the latter's gravitational pull drawing matter from its partner. The intense X-ray emissions from the accretion of matter confirmed the presence of a stellar-mass black hole.

Today, thanks to gravitational wave astronomy, dozens more stellar-mass black holes have been detected. The collisions of these black holes create gravitational waves that ripple through spacetime—events that can be detected across the universe. These discoveries have expanded our understanding of the diversity of stellar-mass black holes and hinted that they may be more abundant than previously thought.

Intermediate-Mass Black Holes: The Missing Link

Between stellar-mass and supermassive black holes lies a mysterious category: intermediate-mass black holes (IMBHs). Weighing hundreds to thousands of times the Sun's mass, IMBHs are harder to detect and remain the most elusive type of black hole. Their existence is important because they could explain how supermassive black holes form.

Evidence for IMBHs is growing. Unusual X-ray sources, stellar motions, and gravitational wave signals point to their presence. In 2020, a team using Hubble and Chandra X-ray Observatory found strong evidence for an intermediate-mass black hole in galaxy 47 Tucanae, within a dense star cluster. This suggests IMBHs may form in globular clusters via collisions and mergers of smaller black holes and stars.

Supermassive Black Holes: Titans of the Cosmos

Supermassive black holes are the true giants of the universe. They weigh millions to billions of times the mass of the Sun and sit at the centres of nearly every large galaxy. Sagittarius A*, at the heart of the Milky Way, is a prime example, with a mass of about four million solar masses.

Supermassive black holes have a profound influence on their host galaxies. By regulating star formation, shaping galactic structure, and powering energetic phenomena such as quasars and relativistic jets, they are essential players in the evolution of the cosmos.

One of the most remarkable examples is the quasar 3C 273, located in the constellation Virgo. Discovered in the 1960s, 3C 273 is powered by a supermassive black hole with a mass exceeding several hundred million suns, producing an extraordinary luminosity that can outshine entire galaxies. Observations of quasars have helped confirm the presence of supermassive black holes and revealed their importance as cosmic engines.

Primordial Black Holes: Time Capsules of the Early Universe

Primordial black holes are perhaps the most intriguing type. Theorised to have formed in the first fractions of a second after the Big Bang, they would have emerged from extreme density fluctuations in the early universe. Their sizes could range from microscopic to thousands of solar masses.

While still hypothetical, primordial black holes could help answer major cosmic questions. Some theories suggest they could account for dark matter, or even serve as seeds for the growth of supermassive black holes. Searches for primordial black holes use gravitational lensing surveys, cosmic microwave background studies, and gravitational wave detectors, but so far the evidence remains tantalisingly incomplete.

Exotic Variants: White Holes, Wormholes, and Beyond

Beyond the established types of black holes, theoretical physics offers exotic possibilities. White holes, the hypothetical time-reverse of black holes, would expel matter rather than absorb it. Wormholes could act as shortcuts through spacetime, connecting distant regions of the universe.

While these concepts remain speculative, they inspire fascinating research and fuel science fiction. The work of physicists such as Kip Thorne and others has shown that these ideas are not purely fantasy—they arise naturally from solutions to Einstein's equations, even if we have yet to find physical evidence for them.

A Cosmic Continuum

The diversity of black holes points to a continuum rather than discrete categories. Stellar-mass black holes may merge to form intermediate-mass black holes, which could evolve into supermassive black holes through cosmic time. Primordial black holes, if they exist, may span the entire range of masses.

Each type of black hole opens a different window into the universe, from the life cycles of stars to the conditions of the early cosmos. They are laboratories for extreme physics, revealing phenomena that cannot be replicated on Earth.

Black Holes as Cosmic Time Capsules

Black holes preserve information about the processes that formed them. The mass, spin, and environment of a black hole carry the story of its origins. In this sense, black holes are time capsules, offering clues to events billions of years in the past. By studying their diversity, scientists can reconstruct the history of the cosmos itself.

Black holes are not merely points of destruction; they are gateways to understanding the grand narrative of the universe. Whether they form from the collapse of massive stars, the mergers of smaller black holes, or the conditions of the early universe, they represent one of the most extraordinary processes in nature—a process that transforms the end of one chapter into the beginning of another.

Chapter 5: Detecting the Invisible

By their very nature, black holes cannot be seen directly. They do not emit light, and their gravitational pull is so intense that even photons—the particles of light—cannot escape once past the event horizon. This makes them invisible to conventional telescopes. Yet, black holes leave powerful fingerprints on their surroundings. For centuries, astronomers have devised clever ways to detect their presence by observing these indirect effects.

Gravitational Influence on Nearby Stars

One of the earliest methods to detect a black hole is to observe how it affects nearby stars. Stars orbiting an invisible mass hint at something massive and unseen exerting gravitational influence. A famous example is Sagittarius A*, the supermassive black hole at the centre of the Milky Way.

For decades, teams of astronomers, such as those led by Reinhard Genzel and Andrea Ghez, have tracked the motion of stars near the galaxy's core. They observed stars moving at astonishing speeds along elliptical orbits, which could only be explained by the presence of a compact object millions of times more massive than the Sun. Their work was so groundbreaking that Genzel and Ghez were awarded the Nobel Prize in Physics in 2020 for their contributions.

This method of detection is especially powerful because it provides information not just about the existence of the black hole but also about its mass and the dynamics of its surroundings.

X-ray Emissions from Accretion Disks

Another important way to detect black holes is by studying the radiation emitted by matter falling toward them. As matter spirals inward, it forms an accretion disk, heated to extreme temperatures by friction and gravitational energy. This process emits vast amounts of radiation, particularly in the X-ray part of the spectrum.

Cygnus X-1 is one of the most famous examples of detection via X-rays. Discovered in the 1960s, it was the first strong candidate for a stellar-mass black hole. Astronomers observed powerful X-ray emissions originating from a region too small and intense to be explained by anything other than a black hole accreting matter from its stellar companion. These observations helped confirm theoretical predictions and marked a turning point in black hole astronomy.

X-ray astronomy is thriving, with satellites like NASA's Chandra and ESA's XMM-Newton providing detailed views of accretion disks and mapping black hole environments across the cosmos.

Gravitational Waves: A New Window

One of the most transformative advances in black hole detection came in 2015 with the first direct detection of gravitational waves. These ripples in spacetime, predicted by Einstein's General Theory of Relativity a century earlier, are generated by massive cosmic events—most dramatically, the collision of black holes.

The Laser Interferometer Gravitational-Wave Observatory (LIGO) in the United States detected waves from the merger of two stellar-mass black holes over a billion light-years away. This historic discovery opened a completely new window for astronomy, allowing scientists to "hear" cosmic events rather than just see them.

Since then, dozens of gravitational wave events have been detected, including mergers of black holes of varying masses and spins. These observations have revealed a previously hidden population of black holes and provided fresh insight into their formation and distribution.

Gravitational wave astronomy is still in its infancy, but it promises to revolutionise our understanding of black holes. Future detectors, such as the European Space Agency's LISA mission, aim to detect gravitational waves from supermassive black holes and even primordial black holes.

The Event Horizon Telescope and Direct Imaging

In 2019, the Event Horizon Telescope (EHT) achieved something remarkable: it produced the first-ever image of a black hole's shadow. The target was M87*, the supermassive black hole at the centre of the galaxy Messier 87. The image showed a glowing ring of light surrounding a dark central shadow—the silhouette of the black hole's event horizon.

This achievement was not the work of a single telescope, but a global network of radio observatories spanning continents. By synchronising their observations through a technique called very-long-baseline interferometry (VLBI), the EHT achieved the resolution equivalent to seeing a grapefruit on the Moon.

The EHT's image confirmed decades of theoretical predictions and opened a new chapter in black hole science. It proved that black holes are not just mathematical curiosities, but physical realities whose shadows we can observe directly.

Gravitational Lensing and Indirect Evidence

Another powerful tool in black hole detection is gravitational lensing. Black holes warp spacetime so strongly that they bend light passing nearby, acting as cosmic magnifying glasses. This effect can produce arcs, rings, or multiple images of background objects.

Gravitational lensing has been used to infer the presence of black holes in distant galaxies and to measure their mass with remarkable precision. In some cases, microlensing events—brief flashes of brightness caused by a black hole passing in front of a star—offer a way to detect isolated black holes that would otherwise remain invisible.

The Future of Detection

The coming decades promise astonishing advances. Next-generation gravitational wave detectors, more powerful X-ray telescopes, and space-based interferometers will push detection capabilities further. Instruments such as the James Webb Space Telescope will probe black holes at unprecedented detail, while missions like the Square Kilometre Array will expand our ability to detect gravitational lensing events.

The future of black hole detection lies in multi-messenger astronomy—combining observations of gravitational waves, electromagnetic radiation, and neutrinos to form a complete picture. This holistic approach will not only discover more black holes but will also reveal new physics and perhaps even clues about the nature of space and time itself.

Chapter 6: Black Holes and the Universe

Black holes are not isolated oddities drifting aimlessly through the cosmos—they are central actors in the grand drama of the universe. Far from being mere destructive sinks, they influence the birth and evolution of galaxies, regulate cosmic structures, and may even hold clues to the universe's deepest mysteries.

The Galactic Architects

Supermassive black holes, with masses millions or even billions of times greater than the Sun, reside at the hearts of most galaxies. Their presence is more than a curious coincidence—it is a fundamental part of how galaxies form and evolve.

In our own Milky Way galaxy, Sagittarius A* sits quietly at the core, yet its gravitational influence shapes the orbits of stars and gas clouds across the galaxy's central bulge. Research shows a striking correlation between the mass of a galaxy's central black hole and the properties of its surrounding stellar population—a relationship known as the M–sigma relation. This suggests that black holes and galaxies grow together, influencing each other's destinies.

Astrophysicist Laura Ferrarese of Rutgers University and David Merritt of Rochester University have explored these links, showing how black holes and galaxies may co-evolve. Their work has deepened our understanding of cosmic structure formation and challenged long-held assumptions that galaxies evolve independently of their central black holes.

Regulating Star Formation

Black holes exert influence far beyond their immediate surroundings. When a supermassive black hole actively accretes matter, it can release tremendous energy through powerful jets and winds. These outflows can travel for thousands of light-years, heating or dispersing the gas in a galaxy.

This process can regulate star formation, either triggering bursts of activity by compressing gas clouds or quenching it by blowing gas away. Observations of active galactic nuclei (AGN) and quasars—galaxies with intensely luminous cores powered by accreting black holes—have shown this dual role in action.

One notable example is the galaxy M87, home to the famous black hole imaged by the Event Horizon Telescope. M87's black hole launches relativistic jets extending thousands of light-years, shaping the galaxy's structure. These jets are more than spectacular—they are fundamental to the galaxy's evolution.

Quasars: Cosmic Beacons

Quasars are among the brightest objects in the universe, powered by supermassive black holes consuming matter at astonishing rates. A quasar can outshine its entire host galaxy, producing more energy than hundreds of billions of suns combined.

The first quasars were discovered in the 1960s, when radio astronomers noticed intense radio emissions from seemingly point-like sources. Optical observations revealed them as faint star-like objects with unusual spectral lines. The realization that these were powered by massive black holes changed astrophysics forever.

Quasars are not just curiosities—they are cosmic laboratories. By studying their light, scientists can learn about the conditions of the early universe, the growth of black holes, and the behaviour of matter under extreme conditions.

Black Holes and Dark Matter

One of the great mysteries of modern astrophysics is dark matter—the invisible substance that makes up about 27% of the universe. Some theories propose that primordial black holes, formed shortly after the Big Bang, could account for at least part of dark matter.

While still speculative, this idea has motivated extensive research. Surveys searching for gravitational lensing events or microlensing caused by unseen massive objects are testing whether primordial black holes exist in significant numbers. Discovering such black holes would not only shed light on dark matter but also reshape our understanding of the early universe.

Cosmic Feedback Loops

The relationship between black holes and the universe is not one-way. Just as black holes influence galaxies, galaxies influence the growth and behaviour of their central black holes. Gas flows, star formation rates, galaxy mergers, and cosmic environments all feed back into the evolution of black holes.

Computer simulations by teams like those at the Illustris and EAGLE projects model these feedback loops, showing how galaxy formation and black hole growth are intertwined in complex cosmic ecosystems. These simulations are essential for testing theories and predicting what future observations might reveal.

Black Holes and the Cosmic Web

The universe is structured like a vast cosmic web, with galaxies, clusters, and filaments of matter stretching across unimaginable distances. Black holes, particularly supermassive ones, are woven deeply into this structure. Their growth and energy influence galaxy clusters and intergalactic medium. Enormous black hole jets, stretching millions of light-years, may shape the cosmos beyond their host galaxies.

Black Holes as Cosmic Probes

Black holes also serve as unique laboratories for fundamental physics. Observing how matter behaves near a black hole can test Einstein's theory of relativity under extreme conditions, probe the nature of spacetime, and search for quantum gravitational effects.

Experiments such as those conducted by the Event Horizon Telescope and gravitational wave observatories give us direct probes into these realms. Black holes, in effect, are not just cosmic curiosities—they are tools that allow us to explore the universe's deepest laws.

The Philosophical Dimension

The role of black holes in the universe also raises profound philosophical questions. Are they the ultimate destroyers, erasing information and defying predictability? Or are they cosmic seeds, fostering new structures and possibilities? Some scientists speculate that black holes could even be gateways to other universes—a tantalising idea that blurs the line between science and philosophy.

Astrophysicist John Archibald Wheeler famously said: *"Black holes have no hair"*—meaning they can be completely described by just a few parameters, like mass, spin, and charge. Yet their influence on the cosmos suggests they are far from simple—they are deeply entwined with the story of existence itself.

Chapter 7: Black Holes and Time

One of the most mind-bending and awe-inspiring aspects of black holes is their effect on time itself. Black holes are not merely extraordinary objects in space; they are extraordinary objects in *time.* They warp it, slow it, and in some cases, may offer glimpses of time travel—at least in theory. The interplay between black holes and time lies at the heart of some of the deepest mysteries in physics.

Einstein's Legacy: Time as a Dimension

Albert Einstein's General Theory of Relativity, published in 1915, forever changed our understanding of time. No longer was time seen as a universal constant flowing at the same rate for everyone—it became part of the fabric of spacetime, stretching and bending in the presence of mass and energy.

Einstein showed that massive objects distort spacetime, slowing down time relative to regions with weaker gravity. This phenomenon, known as **gravitational time dilation**, is one of the most extraordinary predictions of relativity. Near a black hole, the effect is taken to the extreme.

Time Slowing at the Event Horizon

The event horizon of a black hole marks the point of no return—a boundary where gravitational pull is so intense that escape becomes impossible. For an outside observer, an object approaching this horizon would appear to slow down infinitely, never quite crossing it. The falling object itself, however, experiences nothing unusual as it passes through the horizon. This curious difference is a striking example of relativity in action.

The difference between perspectives can be illustrated by a thought experiment: imagine an astronaut, Alex, descending toward a black hole. From Alex's perspective, time proceeds normally, but external observers watching through powerful telescopes would see Alex's movements slow, stretched into an eternal descent. For the outside universe, Alex would appear frozen at the edge of the horizon.

This phenomenon was famously portrayed in Christopher Nolan's film *Interstellar*, where the crew visited a planet near a supermassive black hole. Time dilation meant that hours there equated to years elsewhere—a dramatic but scientifically grounded depiction.

Gravitational Lensing and Time

Black holes also warp the path of light, bending it around them in a phenomenon called **gravitational lensing**. This bending doesn't just distort space—it changes how we perceive time. Light from objects behind a black hole can take different paths, arriving at different times. This creates phenomena such as multiple images of the same object and time delays that astronomers can measure.

Gravitational lensing is not just a curiosity—it is a powerful scientific tool. Researchers use it to map dark matter, study distant galaxies, and test the predictions of general relativity with astonishing precision.

Black Holes and Wormholes

One of the most tantalising ideas in theoretical physics is that black holes might be gateways to other points in spacetime. The concept of a **wormhole**—a bridge between distant regions of space and time—emerges from Einstein's equations of general relativity.

Physicist Kip Thorne, a Nobel laureate, explored this idea in depth. Thorne worked on the mathematics of wormholes and consulted for *Interstellar* to ensure the depiction of black holes and time travel was grounded in real science. While wormholes remain speculative, they spark profound questions: Could a black hole be an entrance to another universe? Could time travel be possible through such a structure?

Even if wormholes cannot exist in practice, their theoretical exploration reveals just how strange black holes make our universe appear. They push physics to its conceptual limits and challenge our deepest intuitions about reality.

The Arrow of Time and Black Holes

Black holes also raise questions about the **arrow of time**—the one-way flow from past to future. In classical physics, time's arrow emerges from the second law of thermodynamics: entropy always increases. Black holes, however, complicate this picture.

Physicist Stephen Hawking proposed that black holes may gradually evaporate through Hawking radiation, releasing information back into the universe. This leads to the famous "black hole information paradox," which asks whether information about matter falling into a black hole is lost forever, or whether it can be recovered. Resolving this paradox is one of the biggest challenges in modern physics and could fundamentally change our understanding of time itself.

Experimental Insights: Measuring Time near Black Holes

Though no human has fallen into a black hole (and for safety, none should attempt it), astronomers have found clever ways to test relativity near massive objects. Observations of stars orbiting Sagittarius A*, the Milky Way's central supermassive black hole, have confirmed that time behaves exactly as predicted by general relativity.

One notable project is the GRAVITY collaboration at the European Southern Observatory. By measuring the orbits of stars close to Sagittarius A*, the team detected effects of gravitational time dilation with unprecedented precision. This is not science fiction—it is real physics being tested in our cosmic backyard.

Philosophical Reflections on Time and Black Holes

The strange effects black holes have on time force us to rethink our most basic assumptions. Time, which we experience as steady and universal, becomes relative and malleable near these cosmic objects.

Some philosophers and physicists even speculate that black holes could connect to a deeper understanding of time—perhaps revealing it as emergent rather than fundamental. This ties into the quest for a theory of quantum gravity, which could unify general relativity and quantum mechanics.

The Future of Time Research

The coming decades promise profound advances in understanding how black holes shape time. Projects like the James Webb Space Telescope and next-generation gravitational wave detectors will allow scientists to observe black holes with unprecedented clarity. These observations may reveal new time-related phenomena, helping answer whether black holes are cosmic endpoints or gateways to something even stranger.

Black holes are not just cosmic curiosities—they are cosmic clocks, warping and stretching time in ways that defy intuition. They invite us to step beyond our everyday experience and confront the deepest questions of reality: What is time? How does it begin? How might it end?

Chapter 8: The End of Black Holes

For centuries, black holes were thought to be eternal—indestructible cosmic graves, swallowing matter and never letting go. Their immense gravity made them seem permanent fixtures of the universe, immune to decay. But in the 1970s, Stephen Hawking revolutionised this understanding, revealing that black holes might have lifespans—and that even they can die. This discovery changed our picture of the cosmos and deepened one of science's most profound mysteries: the fate of black holes, and, by extension, the fate of the universe itself.

The Birth of Hawking Radiation

Stephen Hawking's work combined the worlds of general relativity and quantum mechanics—two pillars of modern physics that rarely speak the same language. In 1974, Hawking applied quantum theory to black holes and discovered something astonishing: black holes are not entirely black. They emit a faint glow of radiation, now known as **Hawking radiation**.

This radiation arises because, at the quantum level, the vacuum of space is never truly empty. Particles and antiparticles constantly pop into existence and annihilate one another in pairs. Near the event horizon of a black hole, gravity can separate these pairs, allowing one particle to escape while the other falls in. The escaping particle becomes Hawking radiation, carrying away energy—and therefore mass—from the black hole.

Hawking's prediction was revolutionary. It suggested that black holes slowly lose mass over time, eventually evaporating entirely. For large black holes, this process would take far longer than the current age of the universe. But for small black holes, evaporation could be significant on cosmic timescales.

The Lifespan of Black Holes

How long does a black hole live? The answer depends on its mass. Supermassive black holes, like Sagittarius A* at the centre of our galaxy, weigh millions or billions of times the Sun's mass and would take far longer than the current age of the universe to evaporate—trillions of years or more.

Stellar-mass black holes, formed from dying stars, also endure for far longer than the current cosmic era. But in the early universe, primordial black holes—if they exist—could be small enough that Hawking radiation would cause them to vanish within billions or even millions of years. Detecting such evaporations could provide evidence of primordial black holes and even shed light on dark matter.

The Black Hole Information Paradox

Hawking's discovery also sparked one of the most heated debates in theoretical physics: the **black hole information paradox.** According to quantum mechanics, information about the physical state of a system can never be destroyed. But if black holes evaporate completely, what happens to the information about the matter they swallowed? Is it lost forever, or somehow preserved in Hawking radiation?

This paradox has profound implications. It challenges our understanding of quantum mechanics, relativity, and the nature of reality itself. Over decades, physicists including Leonard Susskind, Gerard 't Hooft, and Juan Maldacena have proposed various solutions—suggesting that information may be encoded on the event horizon's surface or stored in other dimensions. The debate remains unresolved and is one of the greatest unsolved mysteries in physics.

Cosmic Finality: Black Holes and the End of the Universe

The death of black holes ties directly into the question of the universe's ultimate fate. If Hawking radiation causes all black holes to evaporate, then the universe could enter a final phase far different from what we currently imagine.

Some cosmologists envision a "heat death" of the universe, a state of maximum entropy where all energy is evenly distributed and no work can be performed. In this scenario, after trillions of years, even black holes will vanish, leaving behind a cold, dark expanse. Other models suggest a more dramatic end—perhaps a cosmic "big rip" or a cyclical rebirth of the universe.

Real Observations and Challenges

Hawking radiation is a stunning theoretical insight, but detecting it remains a colossal challenge. The radiation from large black holes is incredibly faint, far outshone by other cosmic sources. Nevertheless, experimental physicists are designing ingenious tests, including laboratory analogues of black holes using ultra-cold atoms, to probe the phenomenon.

In 2019, researchers at the University of Vienna used a system of lasers and sound waves to simulate a black hole's event horizon in a lab, and detected analogue Hawking radiation. While not a direct observation, it provides strong support for Hawking's prediction.

Philosophical Reflections on the End of Black Holes

The idea that even the most powerful objects in the universe can fade away invites deep philosophical reflection. Black holes, long seen as eternal symbols of destruction, might themselves have endings. This humbling insight reminds us that permanence is an illusion in a dynamic universe.

The death of black holes also forces us to confront deeper questions about existence and the nature of reality. If information is preserved in some form, what does that mean for the concept of memory, history, and the laws of physics? And if black holes vanish entirely, what remains at the farthest reaches of time?

The Future of Black Hole Research

The study of black hole evaporation and their ultimate fate is an exciting frontier in physics. Future observatories and experiments, from gravitational wave detectors to space-based telescopes like the James Webb Space Telescope, may bring new insights. Black holes may not just be cosmic vacuum cleaners—they may be cosmic time capsules, holding clues about the birth, evolution, and death of the universe.

Black holes challenge us not only to explore the limits of physics but also to explore the limits of human imagination. Their life cycle—from birth to evaporation—may be one of the most profound stories in science. As we peer into the darkness of a black hole, we confront the nature of time, the limits of knowledge, and the ultimate destiny of the cosmos.

Chapter 9: Black Holes in Popular Culture

Black holes are not just objects of science—they are powerful symbols that capture the imagination of humanity. Their mystery, danger, and mind-bending nature have inspired storytellers, filmmakers, artists, writers, and philosophers for generations. They appear as metaphors for the unknown, as warnings of cosmic destruction, and as gateways to infinite possibilities. In this chapter, we explore how black holes have shaped our culture, how they inspire creativity, and what they reveal about humanity's relationship with the cosmos.

Black Holes as Cultural Symbols

From the moment physicists began talking about "dark stars" in the eighteenth century to the popularisation of the term "black hole" in the 1960s, these cosmic objects have captured the public imagination. They have become symbols of mystery, danger, and the ultimate unknown. For some, black holes represent the end—places of no return where matter, time, and reality itself vanish. For others, they are symbols of discovery, representing the boundaries of human knowledge and the thrill of exploration.

In literature, black holes often serve as metaphors for loss, transformation, and the unknowable. They are used to explore philosophical questions about existence, fate, and the limits of human understanding. For example, science fiction author Arthur C. Clarke used black holes as plot devices to explore time travel and the fate of humanity, while writers such as H.P. Lovecraft used them as symbols of cosmic horror—representing forces far beyond human comprehension.

Black Holes in Film and Television

Few scientific concepts have made such a dramatic leap into popular culture as black holes. One of the most famous cinematic portrayals came in Christopher Nolan's 2014 film *Interstellar*. The movie featured a scientifically accurate depiction of a black hole—named Gargantua—developed in collaboration with physicist Kip Thorne, who won the Nobel Prize for his work on gravitational waves. The depiction of Gargantua's spinning accretion disk, gravitational lensing, and time dilation was not only visually stunning but also grounded in real physics, making it a cultural milestone in science communication.

Black holes have also appeared in countless other films and television series. In *Star Trek*, black holes often serve as plot catalysts for exploration, danger, or strange phenomena. In *Doctor Who*, they are portals to other universes and times, embodying the limitless creativity of science fiction storytelling. Black holes in popular media often blur science and fantasy, but they remain powerful tools for exploring big ideas—about time, space, destiny, and human ambition.

Black Holes in Literature

Science fiction literature has long been fascinated with black holes. Writers have used them to probe questions of physics, cosmology, and the human condition. For example, Gregory Benford's *Timescape* explores the concept of time distortion around black holes, while Dan Simmons' *Hyperion* uses black holes as gateways to alternate realities and metaphysical exploration.

Beyond fiction, non-fiction books about black holes—such as Stephen Hawking's *A Brief History of Time* and Kip Thorne's *The Science of Interstellar*—have brought these cosmic enigmas into popular discourse. These works blend science with storytelling, helping readers grasp complex concepts while inspiring wonder. They reveal that black holes are not just scientific objects, but cultural touchstones that speak to curiosity, mystery, and humanity's desire to know the unknown.

Black Holes in Art and Philosophy

Artists have long been drawn to the imagery of black holes—their strange beauty and terrifying mystery. Paintings, sculptures, and digital art depict black holes as swirling vortexes, portals to other dimensions, or cosmic abysses. This visual fascination mirrors the philosophical allure of black holes: they are objects that challenge our understanding of reality, time, and existence itself.

Philosophers use black holes as a metaphor for the limits of knowledge. They represent boundaries beyond which science cannot currently reach—places where known laws break down. In this sense, black holes are more than astrophysical phenomena; they are symbols of the human condition, reminding us of our curiosity, humility, and the limits of human understanding.

The Impact of Black Holes on Public Imagination

The cultural impact of black holes goes beyond art and literature. They have inspired scientific outreach, public lectures, museum exhibits, and even music. Popular science documentaries—such as *The Universe* and *Cosmos*—have brought black holes into the living rooms of millions, explaining their mysteries in ways both accessible and inspiring. These portrayals remind us that science is not just about equations and experiments; it is about storytelling and wonder.

Black holes also spark philosophical debate. Are they destroyers or creators? Portals or prisons? These questions inspire discussions that bridge science, philosophy, and imagination. Black holes encourage us to think deeply—not only about physics, but about the nature of reality itself.

Black Holes as Inspiration for the Future

Black holes remind us that there is still so much to explore. They challenge the boundaries of science and creativity, inspiring scientists, artists, and dreamers alike. The next decades will likely bring even deeper understanding—and perhaps even direct exploration—of black holes. Projects like the Event Horizon Telescope and future gravitational wave observatories promise to transform our understanding further.

Ultimately, black holes in popular culture reflect humanity's greatest trait: our relentless curiosity. They are not just astronomical objects; they are mirrors of our deepest questions. As symbols, they remind us that the universe is stranger and more wondrous than we can imagine. And they invite us all—scientists, artists, dreamers—to continue looking into the darkness, wondering what lies beyond.

Conclusion: The Mystery Ahead

Black holes are among the most extraordinary—and enigmatic—objects in the cosmos. They stand at the crossroads of science and philosophy, challenging our deepest understanding of reality. They are both terrifying and beautiful, places where the known laws of physics collapse and where the very fabric of space and time is warped beyond comprehension. Black holes are cosmic laboratories, offering a unique window into phenomena that stretch from the smallest scales of quantum mechanics to the largest structures in the universe.

In the past century, our knowledge of black holes has advanced tremendously—from speculative "dark stars" to the first image of a black hole's shadow, and from abstract theory to direct observation of gravitational waves. Yet despite this progress, the mysteries of black holes are far from solved. Many fundamental questions remain: What truly happens inside a singularity, where densities become infinite? How exactly do black holes connect with quantum gravity, the elusive theory uniting general relativity and quantum mechanics? Could they hold the key to understanding the nature of dark matter or dark energy, which together make up nearly 95% of the universe's total content?

Black holes also challenge us to rethink concepts we take for granted. They blur the lines between destruction and creation, time and eternity, existence and disappearance. Are black holes the ultimate cosmic tombs, where matter and information vanish forever—or might they be gateways to other realities, as some speculative theories suggest? These questions touch not only the realm of physics but also of philosophy, raising profound inquiries about the nature of reality, the limits of human knowledge, and humanity's place in the cosmos.

The coming decades promise remarkable advances. New observatories, such as the James Webb Space Telescope, promise to peer deeper into the universe's past and study black holes in unprecedented detail. Advanced gravitational wave detectors will continue to reveal the hidden collisions of black holes across cosmic time, offering insights into their life cycles and the history of the cosmos. Emerging technologies may one day allow us to probe black holes more directly—perhaps even unlocking a practical understanding of how to observe or interact with these objects in ways we cannot yet imagine.

These developments mark the dawn of what may truly be a golden age of black hole science. We are standing on the brink of discoveries that could transform our understanding of the universe in ways as revolutionary as Einstein's theory of relativity. Every new observation, every theoretical breakthrough, pushes us closer to answers—but also to new questions, deepening the mystery.

Black holes remind us that the universe is far stranger, more complex, and more wondrous than we can imagine. They are not simply objects in space; they are profound mysteries, symbols of the boundaries of human knowledge. They invite us to question, to explore, to dream. They remind us that science is not a final answer, but an ongoing journey—a quest to look deeper into the darkness, to challenge what we know, and to embrace the unknown.

Our fascination with black holes is not just a scientific pursuit; it is a reflection of something deeply human. They embody our curiosity, our desire to push beyond limits, and our yearning to understand the universe we inhabit. In contemplating black holes, we are contemplating ourselves—our place in space and time, and the questions that will continue to inspire generations to come.

The mystery of black holes is far from over. As we look to the future, we stand ready for revelations that may reshape science, philosophy, and culture. Black holes will remain gateways—not just to the unknown regions of space, but to the unknown depths of human understanding. And as our tools, knowledge, and imagination grow, so too will the horizon of discovery. Our journey is only beginning.

Glossary of Terms

A quick-reference section for readers to revisit key concepts and terms used in the book.

- **Accretion Disk** – A disk of gas, dust, and stellar debris swirling around a black hole, heated to extreme temperatures as it falls inward.
- **Event Horizon** – The invisible "point of no return" around a black hole where gravity becomes so strong that nothing, not even light, can escape.
- **Ergosphere** – A region outside the event horizon of a rotating black hole where spacetime is dragged along with its rotation.
- **Gravitational Waves** – Ripples in spacetime caused by massive objects in motion, such as merging black holes or neutron stars.
- **Hawking Radiation** – Radiation theorized to be emitted by

black holes due to quantum effects, potentially causing them to lose mass over time.

- **Singularity** – The infinitely dense core of a black hole where spacetime curvature becomes infinite and the laws of physics break down.
- **Wormhole** – A hypothetical passage through spacetime that could connect distant regions of the universe or different universes entirely.

Key Discoveries Timeline

A concise chronology to help readers see how our knowledge of black holes developed.

- **1784** – John Michell proposes the concept of a "dark star" that could trap light.
- **1915** – Albert Einstein publishes the General Theory of Relativity, revolutionizing our understanding of gravity.
- **1916** – Karl Schwarzschild finds the first solution to Einstein's equations describing a black hole.
- **1960s** – John Wheeler coins the term "black hole."
- **1974** – Stephen Hawking predicts black holes can emit radiation.
- **2015** – LIGO detects gravitational waves from colliding black holes.
- **2019** – Event Horizon Telescope produces the first image of a black hole's shadow.

Fun Facts About Black Holes

Engaging tidbits to inspire wonder.

- A teaspoon of matter from a black hole would weigh about **a billion tons** on Earth.
- Black holes can spin at nearly the speed of light, dragging spacetime around them.
- Some black holes can "sing" — producing gravitational waves that scientists can translate into sound.
- The closest known black hole to Earth is about **1,000 light-years away** in the constellation Monoceros.
- The largest known black hole, TON 618, weighs around **66 billion times the mass of our Sun**.

Further Reading & Resources

For curious minds who want to explore beyond this book.

- *Black Holes and Time Warps* by Kip S. Thorne
- *The Black Hole War* by Leonard Susskind
- NASA Black Hole Information Hub: https://www.nasa.gov/black-holes
- Event Horizon Telescope Project: https://eventhorizontelescope.org
- LIGO Gravitational Wave Observatory: https://www.ligo.caltech.edu

Black holes remind us that the universe still holds mysteries beyond imagination — and that the quest to understand them is far from over.

We hope you enjoyed this journey — stay tuned, because the story of black holes is just beginning.

Zac Apos

About the Author

Zac Apos is a passionate explorer of science and discovery, driven by a fascination with the mysteries of the universe. With a rare ability to make complex ideas both vivid and accessible, he inspires curiosity and wonder in readers of all ages.

In *Black Holes: Gateways to the Unknown*, Zac Apos delves into one of the most captivating frontiers of modern science. He unravels the story of black holes — from their mysterious origins and mind-bending anatomy to their powerful role in shaping the cosmos. Drawing on the pioneering work of Einstein, Hawking, and other great minds, Zac reveals why black holes are not merely destructive voids but gateways to understanding the very fabric of reality itself.

www.ingramcontent.com/pod-product-compliance
Lightning Source LLC
LaVergne TN
LVHW101617210926
848480LV00009B/785